Consciousness and the Brain

LAB MANUAL 1.0

EZEQUIEL MORSELLA

© 2022

Preface

This lab manual was intended for students joining a research laboratory on consciousness and the brain. Upon completing it and receiving feedback from students and colleagues, I realized that this little manual could be of interest to a much wider audience, including people outside of academia who are interested in the fascinating topic of consciousness and the brain. I certainly would have benefited from such a manual when I was an advanced undergraduate student!

It is not intended to be a textbook or to present new, original ideas. It is simply a useful compendium that serves as a substitute for assigning new lab members a plethora of research reports and review articles about our research. It presents the information in a streamlined and direct manner.

For readability, conciseness, and clarity, I have omitted almost all of the references. The complete references for all the ideas and observations presented in this manual can be found in Morsella et al. (2016), Morsella and Bargh (2011), and Yankulova et al. (2022). Most of the ideas presented in this book are based on the material presented in these three journal articles, which also contain more in-depth treatments of the phenomena.

5

CONTENTS

I.	Basic Terms	7
II.	Brain Regions	11
III.	Mechanisms	19
IV.	Passive Frame Theory	35
V.	Principles of Operation	41
VI.	Motorium	47
VII.	Practical Tips for the Lab	57
VIII.	About the Author	63
IX.	References	67
X.	Notes	79

Basic Terms

Any particular thing one is conscious of has been referred to as a **CONSCIOUS CONTENT**. A conscious content could be a percept, urge, or autobiographical memory (e.g., memory of one's eighth birthday). The **CONSCIOUS FIELD** is made up of all the conscious contents that are activated at one moment in time (**Fig. 1**).

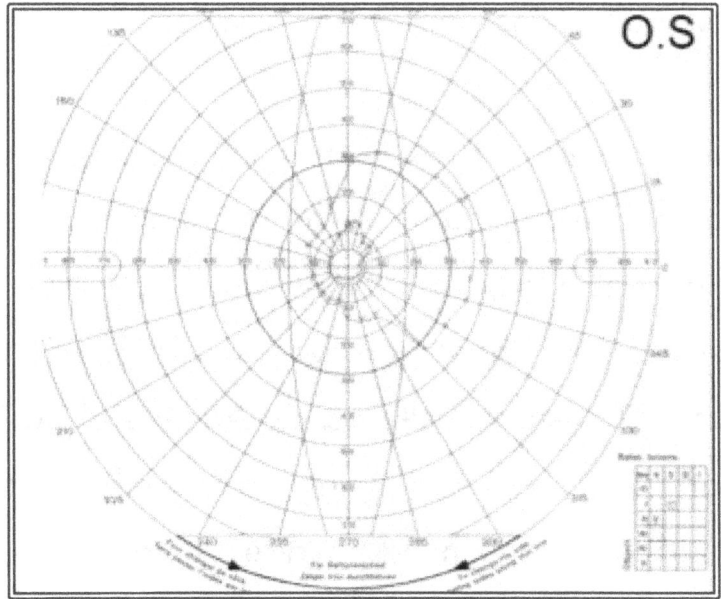

Figure 1. The conscious field is capacious, including memories, urges, and the sensory fields of all the sensory modalities. Depicted here is just one of these fields, perhaps the one that has been studied and mapped out most extensively: the visual field (Source: Wikipedia: Public Domain).

We are here speaking about the most basic form of consciousness: If a creature is capable of having an experience of any kind—pain, nausea, ringing in the ears, a pleasant dream, the sound of a bell, or yellow afterimage—then it possesses this basic form of consciousness. In short, to have an experience of any kind is to be conscious.

Sometimes people refer to this kind of basic consciousness as "awareness." With this term, we could say that, if one is aware of an object before one, the smell of lavender, or aware of an earworm (a song one can't get out of one's head), then one possesses basic consciousness.

In contrast, we are not aware of many things going on in the brain or body—peristalsis in the gut, how the pupils in the eye are controlled, and many other activities in the nervous system. These processes are said to be unconscious. There is usually no experience about them. We know of these processes mainly through reading about them in textbooks. We have no direct experience about them.

For present purposes, *unconscious events are those processes that, though capable of systematically influencing behavior, cognition, motivation, and emotion, do not influence the organism's subjective experience in such a way that the organism can directly detect, understand, or self-report the occurrence or nature of these events* (from Morsella & Bargh, 2011).

Brain Regions

This section presents in a simplified and direct manner thorough reviews of the literature which were presented in Morsella, Krieger, et al. (2010), Morsella and Bargh (2011), Merrick et al. [2014], and Morsella et al. (2016).

Consciousness is associated with only a subset of all of the processes and regions of the nervous system.[1] Many processes in the nervous function are unconscious. These processes include low-level perceptual analysis (e.g., motion detection, color detection, auditory analysis), motor programming, and semantic-conceptual processing. In some cases, the entire processing arc between stimulus-response is mediated unconsciously (e.g., automatisms). In short, action plans can be activated, selected, and even expressed unconsciously.[2]

Some theorists have proposed that, while the cortex may elaborate the contents of consciousness, consciousness is primarily a function of subcortical structures (e.g., Merker, 2007; Penfield & Jasper, 1954; Ward, 2011). This has led to the "cortical-subcortical controversy." The role of subcortical structures in the production of consciousness, and the amount of cortex that may be necessary for the production of consciousness, remains to be elucidated.

Regarding the necessity of the integrity of the frontal lobes for consciousness, it is important to consider that frontal lobotomy, a once common neurosurgical intervention for the treatment of psychiatric disorders, was never reported to render patients incapable of sustaining consciousness.

The olfactory system provides clues regarding the neural underpinnings of conscious perceptual content. The olfactory system is a phylogenetically old system whose circuitry appears to be more tractable and less widespread in the brain than that of other sensory modalities (e.g., vision). Compared to processes such as music appreciation and language perception, it is more "low-level." Several features of the olfactory system render it a fruitful arena in which to isolate the neural underpinnings of consciousness.

First, olfaction involves a primary processing area that consists of paleocortex (which contains only half of the number of layers of neocortex) and primarily only one brain region (the frontal cortex). In contrast, vision and audition often involve large-scale interactions between various brain regions, including the frontal cortex and the parietal cortices.

Second, olfaction can reveal much about the role of thalamic nuclei in the production of consciousness: Unlike most sensory systems, olfactory afferents bypass the first-order, relay thalamus and directly target the cortex ipsilaterally. This direct path minimizes spread of circuitry, which allows one to draw conclusions about the necessity of first-order thalamic relays in (at least) this form of consciousness (i.e., olfactory consciousness).

By investigating the olfactory system, one can also draw conclusions about second-order thalamic relays (e.g., the mediodorsal thalamic nucleus; MDNT). After cortical processing of the olfactory signals, the MDNT receives inputs from olfactory cortical regions. There is no evidence that a lack of olfactory consciousness results from lesions of any kind to the MDNT. We have searched everywhere for such evidence. Regarding second order thalamic relays such as the MDNT, one must keep in mind that, in terms of circuitry, these nuclei are similar in nature to first order relays. Hence, the circuitry in second order thalamic relays is quite simple compared to, say, a cortical column.

In support of 'cortical' theories of consciousness, Cicerone and Tanenbaum (1997) observed complete anosmia (the loss of the sense of smell) in a patient with a lesion to the left orbital gyrus of the frontal lobe. In addition, a patient with a right orbitofrontal cortex (OFC) lesion experienced complete anosmia. This suggests that the OFC is necessary for olfactory consciousness. Consistent with these findings, conscious aspects of odor discrimination have been attributed to the processes of the frontal and orbitofrontal cortices (Buck, 2000). Keller (2011) concludes, "There are reasons to assume that the phenomenal neural correlate of olfactory consciousness is found in the neocortical orbitofrontal cortex" (p. 6; see also Mizobuchi et al., 1999).

According to Barr and Kiernan (1993), olfactory consciousness depends on the piriform cortex. It is worth pointing out, however, that not all lesions of the OFC have resulted in anosmia: Zatorre and Jones-Gotman (1991) reported that OFC lesions yielded severe deficits, yet all the patients observed in the study demonstrated normal olfactory detection.

Another output pathway from the piriform cortex projects to the insular cortex. The insular cortex has anatomical connections to the ventral posteromedial (VPM) nucleus of the thalamus.

Taken together, all of this information (including the conclusions presented above about the MDNT), could lead one to propose that olfactory consciousness depends on the integrity of the insula and thalamus. However, regarding the thalamus, it has been observed that such lesions, including those of the VPM, never result in anosmia. Moreover, regarding the role of the insula in the production of olfactory consciousness, we concur with Mak et al. (2005) that there is no evidence that anosmia results from damage of any kind to the insular cortex.

In conclusion, the foregoing leads one to conclude that the next hypothesis to falsify is that olfactory consciousness requires cortical processes (**see box "Parsimony, Definitions, and Falsifiability"**). This hypothesis is far from obvious, and it is falsifiable, because there are strong frameworks proposing that consciousness is a function of subcortical processes.

Cognitive and Phenomenological Aspects of Olfaction

The olfactory system holds other clues. There are phenomenological and cognitive/mechanistic properties which render the olfactory system a fruitful arena in which to investigate consciousness. Interestingly, unlike what occurs with other sensory modalities (e.g., vision), olfaction regularly yields no subjective experience of any kind when the system is under-stimulated. This occurs when odorants are in low concentration or during sensory habituation. This 'experiential nothingness' is more similar to the phenomenology of the blind spot than to what one experiences when visual stimulation is absent (darkness). The creation of a conscious olfactory content is a true 'addition' to the conscious field—not only does such entry involve the information about a particular stimulus, but it involves the addition, from one moment to the next, of an entire sensory modality.

PARSIMONY, DEFINITIONS, AND FALSIFIABILITY

For a hypothesis to be scientific, it must be capable of being proven false, that is, it must be "falsifiable," a term introduced by Sir Karl Popper. In short, the hypothesis must be able to be tested empirically in some way. Of two hypotheses accounting for the same number of observations, science favors the simpler hypothesis, that is, the one with the fewest assumptions, moving parts, and/or cognitive mechanisms. This is the principle of parsimony.

In science, a complete definition of a phenomenon is the end, and not the beginning, of scientific inquiry. In the beginning of scientific inquiry, one needs only an "identification" or a working, "operational" definition. A complete definition of consciousness is unnecessary to begin to study the phenomenon. (This notion, too, stems from Popper.) The definitions provided in this chapter are useful "identifications."

Mechanisms

After an unexpected nap on a train, one might wake up and experience the sound of the train whistle. The activation of such a conscious content often "just happens." Such "involuntary entry into consciousness" ("involuntary entry," for short) can also stem from a combination of external stimuli and the particular task set that is activated (set-based entry). A set is the disposition to act or think in a certain manner (see box on **"Sets and Consciousness"**).

COMPARE AND CONTRAST

Involuntary entry can be distinguished from forms of entry that are voluntary, such as directed thought, mental rehearsal, working memory use, and mental simulation (see discussion in Jantz et al., 2014; Morsella & Krauss, 2004).

Researchers have begun to investigate the link between external stimuli and involuntary entry using variants of well-known experimental paradigms, including the Stroop task (Morsella, Gray, et al., 2009; Stroop, 1935 **[see box "The Stroop Task"]**), and the flanker task (Eriksen & Eriksen, 1974; Desender et al., 2014; Morsella, Wilson, et al., 2009; Questienne et al., 2018; **[see box "The Flanker Task"]**). More recently, we have been using the reflexive imagery task (RIT), which was developed in the laboratory and is the primary task used in the laboratory.

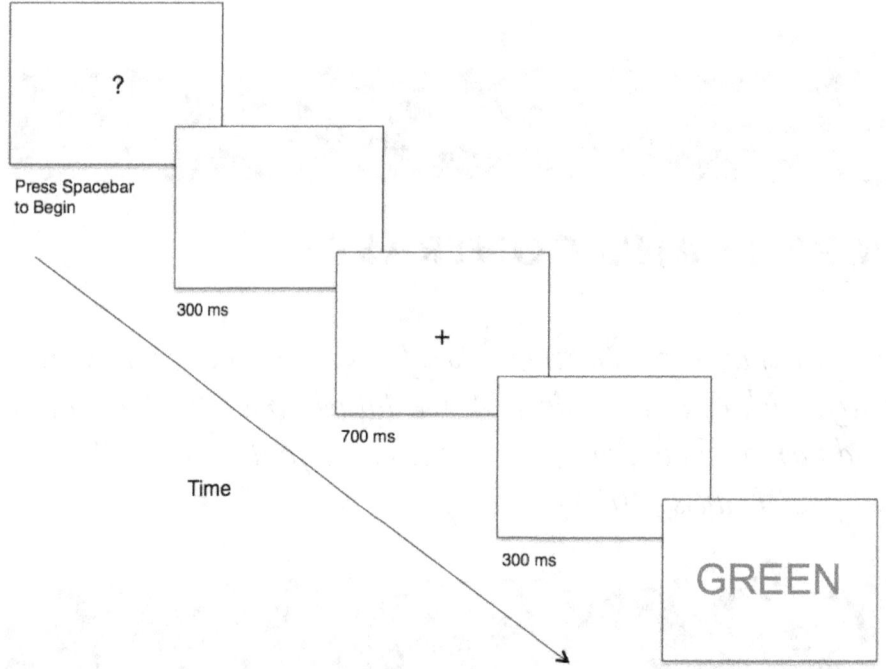

Figure 2: Schematic depiction of an incongruent Stroop trial, with "GREEN" presented in a color other than green (not drawn to scale).

The RIT was developed to investigate set-based entry. It stems from variants of the Eriksen flanker task (e.g., Morsella, Gray, et al., 2009; Morsella, Wilson, et al., 2009), theoretical developments (Morsella et al., 2016) and experimental findings (Ach, 1905/1951; Eriksen & Eriksen, 1974; Gollwitzer, 1999; Morsella et al., 2016; Uznadze, 1966; Wegner, 1989). To fully appreciate this task, one must understand the Stroop and flanker tasks.

THE STROOP TASK

*In the Stroop task, subjects are instructed to name the color in which a word is written (**Fig. 2**). When the word and color are incongruent (e.g., RED presented in blue), "response conflict" leads to increased error rates, response times, and reported urges to make a mistake. (To obtain the "urges to err" measure, after each trial, "How strong was your urge to make a mistake?", which they rate on an 8-point scale, in which 1 signified "almost no urge" and 8 signified "extremely strong urge.") It has been proposed that, in the incongruent condition, there is conflict between word-reading and color-naming plans. When the color matches the word (e.g., RED presented in red), or is presented on* *a neutral stimulus (e.g., a series of x's as in "XXXX"), there is little or no interference.*

The Stroop task possesses a limitation: The incongruent conditions cannot be used to distinguish the effects of interference occurring at different stages of processing (e.g., at perceptual-semantic levels or response selection levels).

Similar to the Stroop task, in the RIT, subjects are instructed not to perform a given mental operation in the presence of certain stimuli. For example, subjects might be instructed to not subvocalize the name of a to-be-presented stimulus (Allen et al., 2013 [**Fig. 3**]).[3] (To subvocalize is to name in one's mind but not aloud.) On a substantive proportion of the trials (on ~80% of the trials; Allen et al., 2013), the RIT effect arises: the stimulus CAT yields the activation of "cat" (i.e., /k/, /æ/, and /t/). The phonological imagery experienced in the RIT is most likely associated with the superior temporal sulcus. The task set (not to perform some mental act) is most likely associated with activations in the frontal cortex. For discussion of the neural correlates of the basic RIT effect, see Dou et al. (2020).

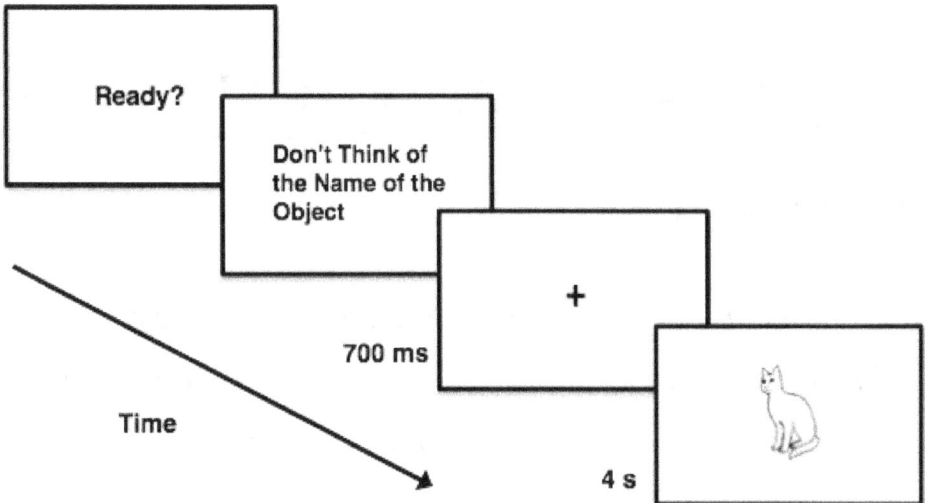

Figure 3: Schematic depiction of an RIT trial (not drawn to scale).

RIT effects have arisen for operations more complex than that of subvocalizing the names of objects. For example, RIT effects have arisen from (a) syntactic processing (Bui et al., 2019); (b) mental rotation (Cushing et al., 2019); (c) musical imagery (White, et al., 2018); (d) high-level shifts in spatial attention (Gardner et al., 2020); (e) the kind of symbol manipulation found in the childhood game of Pig Latin (Cho et al., 2016); (f) insight-related processes (e.g., the insight that "candle" is associated with the stimuli WAX and FLAME; Bui et al., 2019); (g) and the kind of sophisticated visuospatial imagery that occurs in chess (Cushing et al., 2019). For in-depth reviews of the RIT, see Yankulova et al. (2022) and Bhangal, Cho, et al. (2016).

The Flanker Task

*In one version of the task (Eriksen & Schultz, 1979), subjects are trained to press one button with one finger when presented with the letter S or M and to press another button with another finger when presented with the letter P or H. Subjects are then instructed to respond to the stimulus presented in the center of an array (e.g., SSPSS, SSMSS, targets underscored) and to disregard the flanking distractors (**Fig. 4**). In the original flanker task, subjects were instructed to "respond only to the letter in [a] location and to ignore any and all other letters" (Eriksen & Eriksen, 1974, p. 144).*

Response times and self-reported, trial-by-trial 'urges to err' are greater when distractors are associated with a response that is different from that of the target (response interference [RI]; e.g., SSPSS) than when the distractors are different in appearance but associated with the same response (perceptual interference [PI]; e.g., SSMSS), a difference attributed to the automatic activation of response codes by distractors.

Responses are fastest when flankers and targets are identical (e.g., SSSSS). The flanker task reveals, among other things, that introducing interference at different stages of processing (e.g., perceptual versus response selection) leads to distinct behavioral, neural, and subjective effects (see discussion in Morsella, Wilson, et al., 2009).

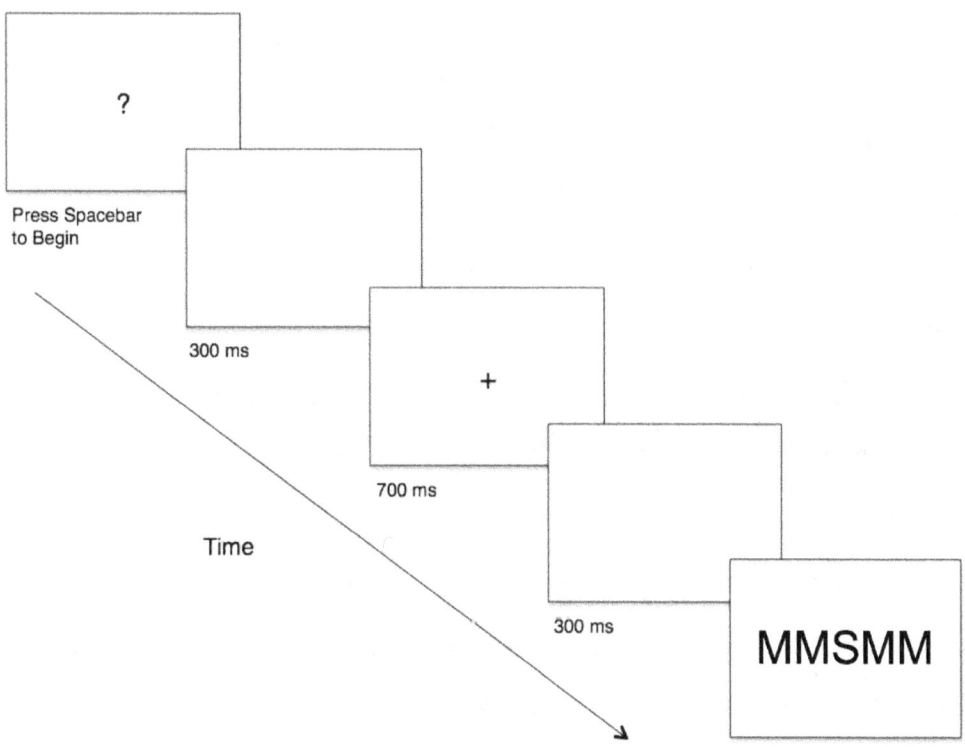

Figure 4: Schematic depiction of a flanker trial (not drawn to scale).

SETS AND CONSCIOUSNESS

In addition to the involuntary entry of percepts and urges, there is also the involuntary entry from the activation of "sets." Sets are the dispositions to act, perceive, or think in certain ways. These sets are associated with the activities of the dorsolateral prefrontal cortex, a region of the brain that is essential for the control of behavior and cognition. Regarding such sets, the theorist Ach noted that, if one activates the set to add, for example, then one cannot help but think of the word "five" after hearing "2 and 3." He referred to sets as "determining tendencies," because they can determine what enters consciousness. At one moment in time, one may not know which sets are activated and will influence consciousness—a form of "imageless thought." This form of involuntary entry could be construed of as set-based entry, which seems to have more moving parts than the involuntary entry of percepts and urges.

Merrick et al. (2015) instructed subjects not to perform two tasks on the stimuli (visual objects): think of the name of the object, and count the number of letters in the object name. Both mental operations occurred involuntarily on ~ 30% of the trials.

Corroboration of subjects' self-reports stem from several sources (see review of evidence in Gardner et al., 2020, Sections 2 & 3). For example, evidence that corroborates that subjects are in fact experiencing the mental imagery stems from the observations that (a) word frequency influences the likelihood and latency of RIT effects (Bhangal et al., 2015), (b) when the involuntary imagery is of counting and of Chess-like spatial imagery, accuracy is high (Bhangal et al., 2018), and (c) the imagery occurs too quickly to be due to strategic processing and can occur under cognitive load (Allen et al., 2013; Cho et al., 2014).

According to all present theoretical accounts, the RIT effect is involuntary (see discussion in Gardner et al., 2020, Sections 2 & 3). Theorists have proposed that RIT effects stem from the "encapsulated" nature of the generation of the majority of conscious contents. Perceptual illusions are said to be encapsulated, because knowledge of the true nature of the perceptual stimulus cannot affect (e.g., turn off) the illusion. The notion of encapsulation (discussed in the next section) is consistent with a recurring idea in the history of psychology—that one is aware of the outputs of mental operations but not of the operations themselves. Theorists such as Helmholtz, Lashley, and George Miller espoused this view. For a historically-based discussion of the origins and merits of the RIT, see Allen et. (2013).

ELIMINATING THE RIT EFFECT

Bhangal, Allen, et al. (2016) concluded that, if RIT effects resemble a reflex, then the effects might habituate the way that reflexes do. Bhangal, Allen, et al. (2016) presented the same stimulus object (e.g., CAT) on ten consecutive trials, to induce habituation (i.e., a weakened RIT effect). From the first presentation of stimulus to its tenth presentation, the mean proportion of trials yielding an RIT effect decreased systematically (SDs in parentheses): .77 (.25), .54 (.32), .46 (.34), .41 (.35), .41 (.36), .40 (.35), .40 (.36), .39 (.37), .39 (.37), .38 (.36). The effect in Bhangal, Allen, et al. (2016), which revealed a negative exponential function, was stimulus-specific: After habituation, the presentation of a different stimulus resulted in the "recovery of the habituated response" (Rankin et al., 2009, p. 137). It is important to consider that, when trying to understand the mechanisms underlying the RIT effect, habituation effects are more informative than interference effects, because the latter can arise in various ways, many of which are trivial.

The RIT provides a technique that can test the limits of involuntary processes without relying on subliminal stimuli. (Subliminal stimuli are visual stimuli that cannot be perceived consciously.) Subliminal stimuli can be problematic because these imperceptible stimuli are not only unconscious, but they are also of weak strength (**see box on "Subliminal Techniques"**).

EVERYDAY EXAMPLE OF INVOLUNTARY ENTRY INTO CONSCIOUSNESS

After a nap on the beach, one might experience the smell of sunblock or the desire to shift one's resting position, and the sight of blimp advertising a cold beverage. The words on the blimp (e.g., Coca-Cola) would activate automatically their phonological forms (/ˈkoʊ.kə ˈkoʊ.lə/). This effect would be a case of visual stimulus eliciting entry of a mental representation that is based on a non-visual modality (audition). It is interesting to note here that, when Helmholtz [1856/1925] discussed unconscious inferences, he was referring not only to basic perception, such as depth perception, but also to the automaticity and insuppressibility of word reading. This example illustrates what often occurs in everyday life—that percepts and urges enter consciousness involuntarily.

Knowledge of the boundary conditions of RIT effects illuminate the limitations of involuntary processes and on the role conscious processes in the control of thought and behavior (discussed in the next section). Of note, the RIT effect will not arise for subliminal stimuli **(see box "Subliminal Techniques"),** nor for operations associated with emotions or autonomic functions (see discussion in Gardner et al., 2020, Section 4). RIT effects will not arise for overt action: subjects are well capable of not uttering aloud the name of objects when instructed not to do so. The effect appears to be associated with the corticospinal tract (also known as the pyramidal tract), which historically has been associated with "voluntary action." This is the topic of the next section.

SUBLIMINAL TECHNIQUES

It is not uncommon in science for 'unconscious processing' to be equated with the kinds of brain processes that occur in response to stimuli that are presented below the threshold of awareness, that is, to stimuli that are presented 'subliminally.' Presenting a stimulus subliminally can be done in several ways, including the technique of 'backward masking,' in which a visual object (e.g., the word POPCORN) is rendered imperceptible by presenting it briefly (e.g., 50 milliseconds) and then having it be followed right away by another visual stimulus (e.g., an array of nonsense symbols such as ######) known as the 'mask.' Substantial research shows that presentation of subliminal stimuli can influence brain processes, behavior, and emotion, at least to some extent.

With this technology, cognitive science began to ask the question, "How smart are responses to subliminal stimuli?" Unfortunately, the data collected to answer this question began to be used as the yardstick by which the sophistication of unconscious processing in general was measured. The data suggested that the unconscious was not that smart. In response to this development in cognitive science and neuroscience, researchers (e.g., Bargh & Morsella, 2008) have cautioned that this is not an accurate way to measure the capacities of the unconscious. Most unconscious processing occurs over, not subliminal stimuli, but stimuli that are consciously-perceptible. In other words, the unconscious usually performs complicated operations that take as their starting point a 'supraliminal' stimulus, that is, something that is consciously perceptible. One is aware of the stimulus but not of its cognitive consequences, events in the brain that influence action and decision making.

Originally appeared as a blog in Psychology Today, 24 December 2010.

34

Passive Frame Theory

Often, external stimuli activate conscious contents in a direct, insuppressible manner. According to Passive Frame Theory (PFT; Morsella et al., 2016), such activations (and the insuppressible nature of the stimulus-elicited effects in the RIT) are advantageous during ontogeny (that is, during development). Consider, for example, that the ability to turn off voluntarily contents such as guilt, nausea, or pain would be detrimental: These contents serve a critical role in guiding behavior, especially during early development. Thus, it has been proposed that the encapsulation of conscious contents, though at times disadvantageous to the actor, is adaptive over the long course of ontogeny.

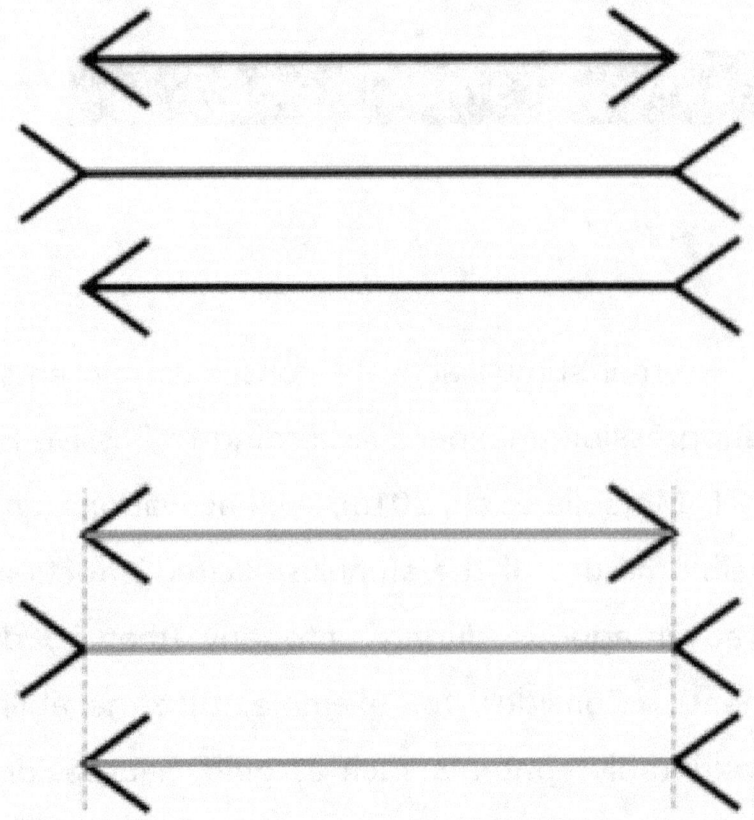

Figure 4. The Müller-Lyer illusion (Source: Wikipedia: Public Domain).

To understand the concept of encapsulation, it is useful to consider the famous Müller-Lyer illusion (**Fig. 4**). The viewer experiencing the illusion is aware, in some sense, that the two horizontal lines are identical in length. Yet, the lines do not seem that way, regardless of one's true knowledge about the illusion. Hence, the illusion is said to be protected or "encapsulated" from the influence of the viewer's knowledge about the stimulus. From this standpoint, visual perception is modular and encapsulated from the rest of cognition, such that perception is "cognitively impenetrable" (Pylyshyn, 1984). Hence, what we perceive is "functionally independent from what and how we think, know, desire, act, and so forth" (Firestone & Scholl, 2016, p. 3).

The conscious contents composing the field cannot influence each other and, figuratively speaking, are unaware of each other and of whether they are action-relevant. These contents function as a lighthouse does: The lighthouse, always in operation, is unaware of which ships can see its light and whether, at that moment, it is serving a critical role.

One analogy that is useful is to think of when trying to understand how the field is constructed is how a song is recorded (Morsella et al., 2022). Each song on the radio is recorded one instrument at a time, on "independent tracks" on a recording console. For example, there is one track for the drums, one track for the guitar, and one track for the vocals. Each of these tracks is independent of the other tracks. In a sense, each is encapsulated from the other tracks. When the tracks are collated in an organized fashion, the song, with all the instruments playing in an organized fashion, is created. The song, with all of the instruments presented simultaneously, is analogous to the conscious field.

The collation in the recording studio involves synchronizing all the tracks temporally. The conscious field is a bit more complex in that it also must organize the conscious contents in terms of their spatial location, with respect to each other and to one (the observer; **see box "The Five Burdens of Encapsulation"**).

According to Morsella et al. (2016), each content is, and should be, encapsulated from the will of the observer and from the influence of the other contents composing the field at that time.

Recently, it has been proposed that conscious contents might be encapsulated from the will of "the observer" as a result of the multidimensional, spatial structure of the conscious field. From this standpoint, the conscious field is a structure in which, according to the rules of projective geometry, the observer must (a) not itself be a conscious content and (b) be separated from all conscious contents (Merker et al., 2022; Rudrauf et al., 2017. **(See box "Projective Geometry.")** This is consistent with the age-old view that the observer cannot directly apprehend, nor introspect about, itself (Schopenhauer, 1818/1819).

PROJECTIVE GEOMETRY

Projective geometry was used by Renaissance painters to create the illusion of depth in their paintings. Regarding our visual sense, notice that the conscious field has three dimensions: one horizontal (the horizon seen from the beach), one vertical (which allows one to see the clouds above the horizon), and one that involves depth (a sand toy near one versus a ship that is far away). According to "viewpoint theory," which is based on the principles of projective geometry, for something to observe these three dimensions (as we do), the origin of the vantage point must reside outside these three dimensions.

Hence, the observer must exist in an extra dimension, which would be, in our case, a fourth spatial dimension. It is proposed that this observer cannot ever directly observe itself, because observing the fourth dimension would then require an additional, fifth spatial dimension. This is called the "n + 1" problem: To apprehend an n number of dimensions, the observer must be in the n + 1 dimension. Viewpoint theory explains an observation noticed by the philosophers Hume and Schopenhauer: That the conscious observer cannot directly apprehend, nor introspect about, itself. It perceives only other things, such as percepts, urges, and ideas. Merker et al. (2022) propose that this principle applies not only to perception but also to the imagination and dream world: One cannot even imagine the nature of a fourth spatial dimension. Our conscious field, and its model of the world, are incapable of representing this dimension in which, in a sense, the conscious observer resides. The conscious observer is not able to directly introspect about itself.

I find it easier to understand this idea by thinking of a simpler case. Imagine that one lived in a two-dimensional world, such as that of a line drawing on a sheet of paper. The line drawing is of a tree. If one lived on the paper, that is, on the planes of this two-dimensional world, one would not ever be able to see the whole drawing on the page. It is only from the third spatial dimension, outside the plane of the two dimensions composing the drawing, that one could see the tree. Similarly, to perceive the three spatial dimensions of our perceptual model of the world, the observer must reside in a fourth spatial dimension, which cannot be perceived in any way.

Originally appeared as a blog in Psychology Today, 7 April 2022.

Principles of Operation

When considering how conscious contents are activated automatically and are encapsulated from each other, the question arises, "How does adaptive, flexible behavior arise from such a rigid system?" According to Passive Frame Theory (PFT; Morsella et al., 2016), which stems from the RIT, encapsulated contents can influence overt behavior collectively, but only through the conscious field. The collective influence from all conscious contents is what yields what appears to be flexible, context-sensitive **ACTION SELECTION**. (Action selection, as when one presses one button versus another button or moves leftwards versus rightwards, is distinct from motor control/motor programming [Proctor & Vu, 2010], processes which are largely unconscious [discussed below].)

Outside the conscious field, the contents can influence behavior, but not collectively. Such unconscious behavior yields 'un-integrated' actions (Morsella & Bargh, 2011). The un-integrated actions can be sophisticated (e.g., the handling of tools in anarchic hand syndrome or in utilization behavior), but they are not influenced by all the types of contextual information by which they should be influenced.

Un-integrated actions lack the kind of context-sensitivity that is obvious in adaptive behavior (e.g., holding one's breath while underwater). For example, in anarchic hand syndrome, the anarchic hand just might grab a cupcake that belongs to someone else or might out of the blue unbutton a button in the sleeve. It is important to note that these actions are not unsophisticated behaviors (neither a robot nor a 3-year-old can unbutton clothing). Rather, these actions are "un-integrated" actions that reflect encapsulation and not a fully operational conscious field.

As is clear in the case of un-integrated actions, it is detrimental for stimulus-specific action plans, though activated to some extent by external stimuli, to directly control action selection. Action selection is adaptive when it takes into account all the current activations, from the stimulus scene, drives, urges, memory, insights, etc. These conscious contents, including the manner in which the spatial locations of these contents are represented (both with respect to each other and to the actor), are essential for adaptive action selection (**see box "Five Burdens of Encapsulation"**). It is for this reason that adaptive action occurs in a manner that is flexible, context-sensitive, and computed on the fly.

According to PFT, action selection must occur in the frame of all the other conscious contents activated at that one instant. This is called a 'frame check' (Morsella et al., 2016). For a successful frame check, the conscious field must operate quickly and be thorough regarding what it represents. In this way, adaptive behavior is not dictated by any single conscious content, but rather by the whole conscious field. The conscious field thus permits the "collective influence" of all conscious contents activated at a given time.

A good analogy for how the conscious field achieves this is a loudspeaker. A loudspeaker can reproduce, and simultaneously present, the many sound waves (frequencies) produced by multiple musical instruments (all through the vibrations of a single diaphragm). Similarly, the conscious field can present, somehow and with great speed and accuracy, a wide variety of conscious contents at one moment in time.

From the present standpoint, consciousness is associated with a stage of processing that involves action options. In line with this theorizing, in one experiment (Bhangal et al., 2018), subjects responded to visual stimuli in one of two ways: by counting the stimuli (Set 1) or naming the color in which they stimuli were presented (Set 2). When selecting one of the two action sets, subjects nonetheless often experienced conscious imagery from the unselected action set.

The conscious field could be construed as a "frame" that affords adaptive action selection, specifically for the the skeletal muscle system, which is the effector system for what in everyday life is called "voluntary behavior" (**see box "Voluntary Action and the Homunculus Fallacy"**). The conscious field itself is passive, like a car window, but essential for encapsulated conscious contents to influence action collectively.

PFT is different from "workspace" models of consciousness (e.g., Baars, 1988; Dehaene, 2014). Those models propose that conscious representations are broadcast to modules that are engaged in both stimulus interpretation and content generation. In PFT, however, the conscious contents are directed only at the unconscious processes of the skeletomotor output system (the topic of the next section). Unlike in workspace approaches, PFT proposes that the conscious field serves only one role. This role is a passive (but necessary) role. In contrast, according to Baars (1988), consciousness serves many functions, including analogy forming, editing and debugging, adaptation and learning, decision-making, metacognitive self-monitoring, and autoprogramming.

VOLUNTARY ACTION AND THE HOMUNCULUS FALLACY

What is a "voluntary" action? Simply stating that an action is voluntary when one intends to do the action is not scientific and introduces the fallacy of the homunculus — that there is a "little man" within the brain, willfully choosing one action plan over another. (The fallacy is that, when introducing the homunculus, one then must also explain how its behaviors and decisions arise. Do they, too, arise from yet another little man inside a mind, this time, inside the tiny mind of the homunculus?)

Scientific approaches to voluntary behavior have stemmed from research on consciousness and action control. Investigations on action control have begun to reveal what conscious states contribute to our behavior. When actions occur unconsciously (e.g., as in neurological disorders or reflexes), they seem to lack "integration," as if the actions are not influenced by all the kinds of information by which they should be influenced. The actions therefore appear impulsive and irrational. In contrast, when actions are mediated by conscious processes, they are "integrated" actions, which are influenced by many sources of information. Integrated behavior occurs, for example, when one is underwater and has the urge to inhale but suppresses this response, or when one is carrying a hot dish of food and experiences the urge to drop the dish and to refrain from dropping it.

Without the conscious field, behavior (including sophisticated motor programming) can arise but the behavior will not be "integrated." For example, in neurological conditions in which actions are decoupled from consciousness and arise involuntarily, sophisticated actions such as manipulating tools or removing clothing can arise but the actions are not influenced by all the kinds of information by which they should be influenced. In a sense, these actions are "un-integrated," appearing insensitive to context.

Originally appeared as a blog in Psychology Today, 20 November 2016. For an in-depth and comprehensive treatment of human action, see Morsella, Bargh, and Gollwitzer (2009), Oxford Handbook of Human Action (Oxford University Press).

Motorium

The conscious field is sampled only by the skeletomotor output system, which is in the service of the somatic nervous system. The somatic nervous system is often contrasted with the autonomic nervous system.

The "response codes" of the skeletomotor output system (once called 'the motorium') are activated inflexibly by the contents of the conscious field. Activate response codes compete with each other over the control of behavior. This competition, and its resolution, is largely unconscious. In the Stroop task, the response codes for "color naming" and "word reading" compete unconsciously.

Returning to the "recording console" analogy, the song one hears on the radio, which is the co-presentation of all the individual tracks, would be analogous to the conscious field, which presents (to the motorium) the unified collation of all the conscious contents activated at one moment in time.

IMPORTANT: In PFT, one kind of unconscious process generates the contents in the conscious field, and another, separate kind of unconscious processing, in the motorium, is activated by these contents. The resolution of the conflict between the "response codes" that are activated (inflexibly) by the conscious contents (e.g., in the Stroop incongruent condition), if such a resolution (representation) exists, is unconscious. In stages of processing, it is a "post-conscious-field" process. Each conflict is idiosyncratic and, if it is to be resolved, requires post-conscious, content-specific algorithms (e.g., one in which overt behavior is influenced most by prepotent action plans; Logan et al. 2015).

If all is working properly **(see box "The Five Burdens of Encapsulation")**, then the conscious field will yield adaptive action selection, which results in "integrated actions" (e.g., holding one's breath while underwater). The conscious field wholly and exclusively determines what in everyday life is called "voluntary behavior" (i.e., integrated actions).

PFT reveals how the kind of reflexive mechanism revealed by the RIT can, when part of a system composed of many such reflexive mechanisms, yield actions that are context-sensitive and more sophisticated than actions from a reflex arc (e.g., the pupillary and patellar reflexes).

THE UNCONSCIOUS MOTORIUM

One is surprised when first learning that most of the sophisticated and intelligent processes in the brain are unconscious. For example, the brain's motor-programs and intersensory processes, as well as the sophisticated actions of the digestive tract, pupils, and respiratory system, and of the 'dorsal pathway' in the brain, are all largely unconscious. These processes are far from dumb or inflexible. When one speaks, one is unconscious of the motor codes telling the lips, jaw, and mouth to move the way that they do. These things are so unconscious that it is often only from reading textbooks about linguistics that one realizes that, regarding what is going on in the mouth, /b/ and /p/ are articulated in the same manner (both are bilabial stops), and so are /d/ and /t/, and /g/ and /k/.

Similarly, one is unconscious of the complicated programs that calculate which muscles should be activated at a given time, but is often aware of their proprioceptive and perceptual consequences (e.g., perceiving a finger flex). It has been proposed by the great William James that action guidance and action knowledge are limited to perceptual-like representations of action outcomes (e.g., the 'image' of one's finger flexing), with the motor programs/events actually responsible for enacting the actions being unconscious. (Motor programming is largely unconscious; see Rosenbaum, 2002.) Similar accounts have been proposed by contemporary action researchers (e.g., Greenwald, Prinz, and Hommel). Interestingly, it is these perceptual-like representation that constitute that which is consciously accessible in normal action, dreams, and when observing the actions of others.

For example, it is the phonological representation (and not, say, the motor-related, articulatory code) that one is conscious of during both spoken and subvocalized speech (when one speaks in one's head), or when perceiving the speech of others. James proposed that the conscious mind later uses these conscious perceptual-like representations to voluntarily guide the generation of motor efference, which itself is an unconscious process. But all the real work and heavy lifting of catching a ball--by getting these but not those muscles to contract--is largely unconscious. If these unconscious mechanisms are so intelligent and capable, then why is it necessary for us to be consciously aware of any process in the brain? One answer is that consciousness is not for motor control but for a higher level of action control. **(This section initially appeared as a blog in *Psychology Today*, 5 February 2010. In-depth treatment can be found in Morsella, Bargh, and Gollwitzer, 2009, *Oxford Handbook of Human Action*. New York: Oxford University Press.)**

IDEOMOTOR THEORY AND UNCONSCIOUS MOTOR PROGRAMMING

The television show *60 Minutes* presented a story about how patients can today control robotic arm/limb prostheses. In the episode, the interviewer was surprised to learn that a soldier who had tragically lost his lower arm in combat could, in just a few trials, control the grasping motions of a robotic hand. This prosthesis was connected to electrodes attached to the muscles of the remaining part of the soldier's upper arm. The interviewer asked the soldier how he knew which muscles to activate in order to enact the robot's action. The soldier replied to the effect that he had no idea regarding which muscles to activate, nor what the muscles were actually doing. Rather, the soldier claimed that, to enact the action on the part of the robotic arm, all he had to do was imagine the grasping action. This image, what Harleß in the nineteenth-century called in German the Effektbild (in English, "the picture or image of the effect") was somehow translated (unconsciously) into the kind of muscular activation that would normally result in a grasping action.

Is this how people generally guide and perceive their own actions? According to William James', the founder of American psychology and popularizer of this kind of ideomotor theory, the answer is a resounding yes. Ideomotor theory states that action guidance, and action knowledge, are limited to perceptual-like representations of action outcomes (e.g., the 'image' of one's finger flexing), with the motor programs/events actually responsible for enacting the actions being unconscious. From this standpoint, conscious contents regarding an ongoing action are primarily of the perceptual consequences of that action. In James' own words, "In perfectly simple voluntary acts there is nothing else in the mind but the kinesthetic idea...of what the act is to be" (*Principles of Psychology*, p. 771). Today, ideomotor theory may prove to be helpful not only for understanding the control of everyday behavior but also for illuminating how patients can benefit from the specific technology mentioned on *60 Minutes* and from the more general forms of brain computer interfaces (BCIs), which must be intuitive and easy to use.

(Originally published as a blog in Psychology Today, 22 April 2012.)

THE FIVE BURDENS OF ENCAPSULATION

Because of encapsulation, each "conscious content" in the conscious field does not in a sense "know" of the nature of the other conscious contents composing the field. Because of encapsulation, each content in the conscious field also does not know whether it is relevant to current goals and actions. In order for a system featuring encapsulation to yield adaptive behavior, several conditions must be met. These conditions could be construed as the five burdens of encapsulation.

First, because no content knows whether it is action-relevant or not, and also does not know of the nature of the other contents composing the conscious field, the conscious field must be very thorough and represent as many (potentially actionable) contents as possible, just in case. This explains why, even though the conscious field is for adaptive behavior, one is often aware of things to which one does not need to respond. Thus, encapsulation explains why the field is so capacious and inclusive.

Second, in order to benefit action selection, each content (e.g., the color blue versus the smell of lavender) must differentiate itself from all the other contents in the field, for a contrast not apparent in the field cannot be reflected in voluntary action. Each content must differentiate itself not only from contents within the same modality (vision), but also from contents from other modalities (smell). These contrasts must arise even though all content must somehow exist in the same decision space, and therefore share the same underlying format.

Third, because of encapsulation, the spatial layout of the stimulus scene must represent spatial coordinates as thoroughly as possible. This occurs for many sensory modalities (but not olfaction). This is because it is often the case that the "discriminative stimulus" that determines which action should be performed is not a single stimulus, but rather the spatial distance between two stimuli, as in the case of driving. Thus, our conscious field must have a rich and thorough representation of the spatial dimensions of the external world. The field does not know which such spatial relation might be essential for adaptive action selection.

Fourth, for action to be adaptive, such a spatial model of the world must include the emergence of first-person perspective, due to the demands of action selection, for example, when deciding between reaching for a large (but faraway) banana on one's right or a smaller (but nearby) banana on one's left. The first-person perspective is essential for this kind of action selection. (This first-person perspective also emerges in the dream world.)

Fifth, because of encapsulation, and in order for action to be adaptive, the contents that compose the conscious field must all be comparable at some level, for they must exist as comparable tokens in a common decision space. These contents include information about the immediate environment, the representations of anticipated actions (e.g., mental imagery of to-be-produced actions), the effects of actual action (e.g., proprioceptive feedback), and even high-level cognitions. All of these contents, which tend to have a perceptual-like format, are sampled not by other conscious contents (which would violate encapsulation), but rather by the action systems in the Skeletal Muscle Output System. These systems are unconscious. In short, encapsulation explains why the conscious field, though in the service of adaptive action, contains contents that are not action-relevant, and why it has a first-person perspective and is so thorough (both in terms of its contents and the representation of spatial coordinates).

Originally appeared as a blog in Psychology Today, 20 December 2018. In-depth treatment can be found in Morsella et al. (2020).

Practical Tips for the Lab

Thanks to the guidance of my undergraduate mentor, Prof. Robert B. Tallarico, when I was an undergraduate student I had the great treat of reading The Evolution of Physics: The Growth of Ideas from Early Concepts to Relativity and Quanta (1938) *by Albert Einstein and Leopold Infeld. Each time I revisit that book, I learn something new and realize how many interesting things I missed during my previous readings. One thing I realized during my very first reading of that book is that, whenever Einstein and Infeld introduced a great insight into the nature of all things (e.g., the nature of light and energy), that insight usually stemmed from the intensive study of a specific, 'strange scenario,' one which is rare and somewhat contrived (e.g., how charged particles are emitted from a zinc plate that is exposed to ultraviolet light).*

I realized that the biggest ideas about many things often came from the repeated study of the same small event, over and over, as in the case of Galileo and his inclined planes. From the intensive study of very few such scenarios seemed to stem many great insights. Picking the right 'strange scenario' to study seemed to be essential for the generation of new insights. Theorizing did not seem to occur in a vacuum but was always tied to very specific, concrete phenomena, phenomena that are experimentally tractable, reliable, and not messy (e.g., having few variables).

The same appears to be the case in neuroscience and psychology. Insights about everyday behavior (in uncontrolled environments) often come from what occurs during a single trial of a highly-controlled laboratory task, such as the classic Stroop task in which one has to name the colors in which words are written and in which performance suffers when the word and the color mismatch (e.g., the word GREEN presented in blue). Many insights about the mind and brain have stemmed from tasks such as the Stroop task, which is a very specific, contrived, and strange circumstance: One does not walk through Times Square naming the colors in which the words in billboards are presented. In most laboratory tasks, the subject responds to a stimulus, and the experimenter controls the nature of the stimulus and, critically, when that stimulus is presented. This is because it is difficult to draw conclusions about the subject's ongoing behavioral and mental phenomena when the phenomena are not tied to a specific stimulus, whose nature is well known, and when the phenomena arise under conditions that are less controlled than those of laboratory tasks.

It is for this reason that I speculate that, when one day there is an explanation concerning how consciousness emerges from brain function, that explanation will stem from observations of a simple scenario in which a human being responds to an external stimulus. My hunch is that the explanation will not concern that kind of consciousness we often experience in everyday life when the contents of the mind wander and are not linked so tightly to the stimuli composing the external world, the world whose true nature perplexes the physicist.

(Originally appeared as a blog in Psychology Today, 28 June 2019.)

And here are some, more practical tips:

In general, in an experiment, never present the same stimulus twice.

Always strive for a within-subjects (repeated measures) design that lasts less than one hour and has the trials representing the various conditions intermixed and presented randomly, as occurs in the Stroop task and flanker task. (Within-subjects designs are more sensitive than between-subjects designs.) For example, a random trial sequence in the Stroop task might be CONGRUENT, INCONGRUENT, INCONGRUENT, CONGRUENT, INCONGRUENT...

If you must block the trials per condition (and there are many reasons for doing so), be sure to counterbalance fully the order of presentation of the two blocks of trials across subjects. For example, Subject 1 will perform Block A first, and Subject 2 will perform Block B first. If there are two blocks, then half of the subjects should perform one sequence (A then B), and the other half of the subjects should perform the other sequence (B then A). In this way, one diminishes the influence of order effects on the dependent variable. If each condition involves a different task set, then it is usually better to use a blocking strategy, as it is difficult for a subject to change the task set on a trial-by-trial basis.

When running subjects, the experimenter must always be present moments (10-15 minutes) before the beginning of the experimental session, so that everything will be ready without delay. Make sure that all of the equipment (e.g., the button box) is functioning properly BEFORE the subject arrives. If you collect data of any kind, get IRB approval beforehand.

Learn how to write a method section by studying the most recent method sections from the laboratory. The most recent method sections reveal the current, constructive demands from the reviewers.

When running a subject, the research assistant always follows the scripted procedure. Deviation from it will introduce variability into the data set and may compromise the findings.

In many experiments, explain to subjects that they must sit in the appropriate position and maintain the same distance from the computer or microphone. In general, subjects should always fixate on the center of the screen. Also, their fingers should always rest on the buttons and should never be lifted, even when pressing, for moving the finger in the air introduces variability into the data set. They must respond as fast and as accurately as possible, but never before the stimulus appears.

Never play around with someone else's script or data without invitation or tamper with the industry-standard computer or psychophysiology equipment

Write the method section as you do the study, so that nothing about the method is forgotten

When in doubt, have everything in your design be BALANCED or EQUAL.

Experiment 2 should be identical to Experiment 1 except for one change, one which is necessary to rule out alternative hypotheses for Experiment 1.

Always have at least 40 stimuli per condition.

Run yourself in your study and make sure the data file has all the info you will need for analysis (this error should not be discovered after running 50 subjects!).

When creativity/productivity is slow, organize files, review the literature, and prepare for the next wave of creativity/productivity.

Always back up your new data and never submit something without the lab director having reviewed it and approved it first. Always lock and alarm the lab.

Always remember that in science, the trick is to enjoy the process!

About the Author

A theoretician and experimentalist in neuroscience, Ezequiel Morsella received his Ph.D. at Columbia University (2002) and carried out his postdoctoral training (2003 - 2007) at Yale University. Since his pre-college days, he has been focusing on the contrast between the conscious (e.g., urges and voluntary action) and unconscious brain mechanisms in human action control, thanks in large part to stumbling across a pile of old books by Hebb and Hull. All of his publications, experimental paradigms, and theoretical developments center around the theme of the nature of involuntary (versus voluntary) "entry into consciousness" during action control.

In 2007, he was hired as a professor in neuroscience at San Francisco State University (where he is now Professor of Neuroscience) and as an Assistant Adjunct Professor in the Department of Neurology at the University of California, San Francisco. His theory has appeared in Psychological Review and Behavioral and Brain Sciences (target article). The research was covered by *TIME Magazine*. His current research has been supported by the Toyota Motor Corporation.

Regarding his training, while at Columbia, he was mentored by Robert Krauss (Morsella's primary advisor, with whom Morsella conducted much research on action control, working memory, and mental representation), and by the cognitive neuropsychologist Michele Miozzo, with whom Morsella studied about models of action/speech production. At Yale, he was mentored by John Bargh and the neuroscientist Jeremy Gray. As an undergraduate, he was mentored by Robert B. Tallarico at the University of Miami.

Today, he is the lead author of *Oxford Handbook of Human Action*. His research has appeared in journals such as *Psychological Review*, *Behavioral and Brain Sciences*, *Perspectives on Psychological Science*, *Neurocase*, *Consciousness and Cognition*, *Experimental Brain Research*, and *Journal of Experimental Psychology: Learning, Memory, and Cognition*. He has served as an editorial reviewer for many journals, including *Science*, *Behavioral and Brain Sciences*, *Journal of Cognitive Neuroscience*, *Psychological Review; Cognition*, *Emotion*, and *Journal of Experimental Psychology (General, HPP, LMC)*.

Concerning how he became interested in science, Morsella writes:

When I was little, I was surprised that lizards could behave so intelligently, even though their brains and bodies are so small. Their eyes, always keeping an eye on me no matter where I went in my Floridian backyard, would move around and track me, seeming more alert--even more 'alive'--than our own human eyes often do. I was even more surprised (shocked, actually) when I saw that a lizard's tail, when detached from the lizard's torso, moves about vigorously, jumping around and looking like a living thing. This unfortunate event for both the lizard and the catcher happens when one tries to catch a lizard from behind but, because of the lizard's speed, ends up empty-handed, with one's fingers failing to reach past the hind legs, making contact only with the tail.

The grown-ups around me explained to me that the detached tail (which grows back, they assured me) moves because it has some residual "life force" from the lizard's body. Years later I learned the truth--that the tail moves because of the activities of excitable nerve cells (neurons) contained within it. These neurons wiggle the detached tail so that a predator becomes enthralled by its movements and forgets about the more important parts of the lizard (the brains, lungs, gametes) that are running away at that moment. Greater was the realization that, whatever makes the detached tail move also makes the lizard's brain work they way it does, giving rise to, for example, its intelligent eye movements. More shockingly was the acknowledgment that the same kind of cells make my own brain work the way it does, with its perceptions, memories, desires, and conscious experiences.

How the kinds of unconscious and unintelligent cells in a lizard tail can ever give rise to our intelligent, conscious mind is one of the greatest mysteries in science, one that I have devoted my life to as a research scientist. Had I been raised in a colder climate, perhaps I would have never had this reptilian experience, one which continues to enthrall my mind today and left a lizard tail-less (momentarily).

Originally appeared as a blog in Psychology Today, 2 September 2009.

References

Ach, N. (1905/1951). Determining tendencies: Awareness. In D. Rapaport (Ed.), *Organization and pathology of thought* (pp. 15 - 38), New York: Columbia University Press. (Original work published in 1905.)

Allen, A. K., Wilkins, K., Gazzaley, A., & Morsella, E. (2013). Conscious thoughts from reflex-like processes: A new experimental paradigm for consciousness research. *Consciousness and Cognition, 22,* 1318 - 1331.

Baars, B. J. (1988). *A cognitive theory of consciousness.* Cambridge, England: Cambridge University Press.

Bargh, J. A., & Morsella, E. (2008). The unconscious mind. *Perspectives on Psychological Science, 3,* 73-79.

Barr, M. L., & Kiernan, J. A. (1993). *The human nervous system: An anatomical viewpoint* (6th ed.). Philadelphia: Lippincott.

Bhangal, S., Allen, A. K, Geisler, M. W., & Morsella, E. (2016). Conscious contents as reflexive processes: Evidence from the habituation of high-level cognitions. *Consciousness and Cognition, 41,* 177-188.

Bhangal, S., Cho, H., Geisler, M. W., & Morsella, E. (2016). The prospective nature of voluntary action: Insights from the reflexive imagery task. *Review of General Psychology, 20,* 101-117.

Bhangal, S., Merrick, C., Cho, H., & Morsella, E. (2018). Involuntary entry into consciousness from the activation of sets: Object counting and color naming. *Frontiers in Psychology.* doi: 10.3389/fpsyg.2018.01017

Buck, L. B. (2000). Smell and taste: The chemical senses (pp. 625–647). In E. R. Kandel, J. H. Schwartz, & T. M. Jessell (Eds.), *Principles of neural science* (4th ed.). New York: McGraw-Hill.

Bui, N-C. T., Ghafur, R. D., Yankulova, J., & Morsella, E. (2019). Stimulus-elicited involuntary insights and syntactic processing. *Psychology of Consciousness: Theory, Research, and Practice,* doi:10.1037/cns0000208

Bush, G., Shin, L. M., Holmes, J., Rosen, B. R., & Vogt, B. A. (2003). The multi-source interference task: Validation study with fMRI in individual subjects. *Molecular Psychiatry, 8,* 60 – 70.

Cho, H., Godwin, C. A., Geisler, M. W., & Morsella, E. (2014). Internally generated conscious contents: Interactions between sustained mental imagery and involuntary subvocalizations. *Frontiers in Psychology, 5.* https://doi.org/10.3389/fpsyg.2014.01445.

Cho, H., Zarolia, P., Gazzaley, A., & Morsella, E. (2016). Involuntary symbol manipulation (Pig Latin) from external control: Implications for thought suppression. *Acta Psychologica, 166,* 37-41.

Cicerone, K. D., & Tanenbaum, L. N. (1997). Disturbance of social cognition after traumatic orbitofrontal brain injury. *Archives of Clinical Neuropsychology, 12,* 173 - 188.

Cushing, D., Gazzaley, A., & Morsella, E. (2019). Involuntary mental rotation and visuospatial imagery from external control. *Consciousness and Cognition, 75.* doi: 10.1016/j.concog.019.102809.Epub 2019 Sep 12.

Dehaene, S. (2014). *Consciousness and the brain: Deciphering how the brain codes our thoughts.* New York: Viking.

Desender, K., van Opstal, F. V., & van den Bussche, E. (2014). Feeling the conflict: The crucial role of conflict experience in adaptation. *Psychological Science, 25,* 675 - 683.

Dou, W., Allen, A. K., Cho, H., Bhangal, S., Cook, A. J., Morsella, E., & Geisler, M. W. (2020). EEG correlates of involuntary cognitions in the reflexive imagery task. *Frontiers in Psychology,* doi: 10.3389/fpsyg.2020.00482.

Eriksen, B. A., & Eriksen, C. W. (1974). Effects of noise letters upon the identification of a target letter in a nonsearch task. *Perception and Psychophysics, 16,* 143 - 149.

Eriksen, C. W., & Schultz, D. W. (1979). Information processing in visual search: A continuous flow conception and experimental results. *Perception and Psychophysics, 25,* 249 - 263.

Firestone, C., & Scholl, B. J. (2016). Cognition does not affect perception: Evaluating the evidence for 'top-down' effects. *Behavioral and Brain Sciences* [Target Article], 39, 1-77.

Gardner, K., Walker, E. B., Li, Y., Gazzaley, A., & Morsella, E. (2020). Involuntary attentional shifts as a function of set and processing fluency. *Acta Psychologica, 203.* doi: 10.1016/j.actpsy.2020.103009.

Gollwitzer, P. M. (1999). Implementation intentions: Strong effects of simple plans. *American Psychologist, 54,* 493 - 503.

Helmholtz, H. v. (1856/1925). *Treatise of physiological optics: Concerning the perceptions in general.* In T. Shipley (Ed.), Classics in psychology (pp. 79 – 127). New York: Philosophy Library.

Jantz, T. K., Tomory, J. J., Merrick, C., Cooper, S., Gazzaley, A., & Morsella, E. (2014). Subjective aspects of working memory performance: Memoranda-related imagery. *Consciousness and Cognition, 25,* 88-100.

Keller, A. (2011). Attention and olfactory consciousness. *Frontiers in Psychology, 2,* Article 380, 1 - 11.

Logan, G. D., Yamaguchi, M., Schall, J. D. & Palmeri, T. J. (2015). Inhibitory control in mind and brain 2.0: Blocked-input models of saccadic countermanding. *Psychological Review, 122,* 115–47.

Mak, Y. E., Simmons, K. B., Gitelman, D. R., & Small, D. M. (2005). Taste and olfactory intensity perception changes following left insular stroke. *Behavioral Neuroscience, 119,* 1693 - 1700.

Merker, B. (2007). Consciousness without a cerebral cortex: A challenge for neuroscience and medicine. *Behavioral and Brain Sciences, 30,* 63 - 134.

Merker, B., Williford, K., & Rudrauf, D. (2022). The integrated information theory of consciousness: A case of mistaken identity. *Behavioral and Brain Sciences, 45,* E41.

Merrick, C., Farnia, M., Jantz, T. K., Gazzaley, A., & Morsella, E. (2015). External control of the stream of consciousness: Stimulus-based effects on involuntary thought sequences. *Consciousness and Cognition, 33,* 217-225.

Merrick, C., Godwin, C. A., Geisler, M. W., & Morsella, E. (2014). The olfactory system as the gateway to the neural correlates of consciousness. *Frontiers in Psychology, 4,* 1011. doi: 10.3389/fpsyg.2013.01011

Mizobuchi, M., Ito N., Tanaka, C., Sako, K., Sumi, Y. & Sasaki, T. (1999). Unidirectional olfactory hallucination associated with ipsilateral unruptured intracranial aneurysm. *Epilepsia, 40,* 516-519.

Morsella, E., & Bargh, J. A. (2011). Unconscious action tendencies: Sources of 'un-integrated' action. In J. T. Cacioppo & J. Decety (Eds.), *The Oxford handbook of social neuroscience* (pp. 335 - 347). New York: Oxford University Press.

Morsella, E., Velasquez, A. G., Yankulova, J. K., Li, Y., Wong, C. Y., & Lambert, D. (2020). Motor cognition: The role of sentience in perception-and-action. *Kinesiology Review, 9,* 261-274

Morsella, E., & Krauss, R. M. (2004). The role of gestures in spatial working memory and speech. *American Journal of Psychology, 117,* 411-424.

Morsella, E., Godwin, C. A., Jantz, T. K., Krieger, S. C., & Gazzaley, A. (2016). Homing in on consciousness in the nervous system: An action-based synthesis. *Behavioral and Brain Sciences* [Target Article], 39, 1-17.

Morsella, E., Gray, J. R., Krieger, S. C., & Bargh, J. A. (2009). The essence of conscious conflict: Subjective effects of sustaining incompatible intentions. *Emotion, 9*, 717-728.

Morsella, E., Krieger, S. C., & Bargh, J. A. (2010). Minimal neuroanatomy for a conscious brain: Homing in on the networks constituting consciousness. *Neural Networks, 23*, 14-15.

Morsella, E., Velasquez, A. G., Yankulova, J. K., Li, Y., & Gazzaley, A. (2022). Encapsulation and subjectivity from the standpoint of viewpoint theory. *Behavioral and Brain Sciences, 45*, 37-38.

Morsella, E., Velasquez, A. G., Yankulova, J. K., Li, Y., Wong, C. Y., & Lambert, D. (2020). Motor cognition: The role of sentience in perception-and-action. *Kinesiology Review, 9*, 261-274

Morsella, E., Wilson, L. E., Berger, C. C., Honhongva, M., Gazzaley, A., & Bargh, J. A. (2009). Subjective aspects of cognitive control at different stages of processing (7-experiment article). *Attention, Perception, & Psychophysics, 71*, 1807-1824.

Penfield, W. & Jasper, H. H. (1954). *Epilepsy and the functional anatomy of the human brain.* New York: Little, Brown.

Proctor, R. W., & Vu, K.-P. L. (2010). Action selection. In I. B. Weiner & E. Craighead (Eds.), *The Corsini encyclopedia of psychology* (Vo. 1, pp. 20 – 22). Hoboken, NJ: John Wiley.

Pylyshyn, Z. W. (1984). *Computation and cognition: Toward a foundation for cognitive science.* Cambridge, MA: MIT Press.

Questienne, L., Atas, A., Burle, B., & Gevers, W. (2018). Objectifying the subjective: Building blocks of metacognitive experiences in conflict tasks. *Journal of Experimental Psychology: General, 147,* 125-131.

Rankin, C. H., Abrams, T., Barry, R. J., Bhatnagar, S., Clayton, D., Colombo, J., … Thompson, R. F. (2009). Habituation revisited: An updated and revised description of the behavioral characteristics of habituation. *Neurobiology of Learning and Memory, 92,* 135-138.

Rosenbaum, D. A. (2002). Motor control. In H. Pashler (Series Ed.) & S. Yantis (Vol. Ed.), *Stevens' handbook of experimental psychology: Vol. 1. Sensation and perception* (3rd ed., pp. 315 - 339). New York, NY: Wiley.

Rudrauf, D., Bennequin, D., Granic, I., Landini, G., Friston, K., & Williford, K. (2017). A mathematical model of embodied consciousness. *Journal of Theoretical Biology, 428,* 106–131.

Schopenhauer, A. (1818/1819). *The world as will and representation, volume 1.* New York: Dover.

Uznadze, D. (1966). *The psychology of set.* New York: Consultants Bureau.

Velasquez, A. G., Gazzaley, A., Toyoda, H., Ziegler, D. A., & Morsella, E. (2021). The generation of involuntary mental imagery in an ecologically-valid task. *Frontiers in Psychology, 12,* 759685. doi: 10.3389/fpsyg.2021.759685

Ward, L. M. (2011). The thalamic dynamic core theory of conscious experience. *Consciousness and Cognition, 20,* 464-486.

Wegner, D. M. (1989). *White bears and other unwanted thoughts.* New York: Viking/Penguin.

White, N. A., Velasquez, A. G., & Morsella, E. (2018). *Mary had a little… : Involuntary music imagery and memory retrieval.* Poster presented at the Annual Convention of the Association for Psychological Science, San Francisco.

Yankulova, J. K., Zacher, L. M., Velasquz, A. G., Dou, W., & Morsella, E. (in press). Irrepressible cognition in the reflexive imagery task: Insights and future directions. *Frontiers in Psychology: Cognition.*

Zatorre, R. J. & Jones-Gotman, M. (1991). Human olfactory discrimination after unilateral frontal or temporal lobectomy. *Brain, 114,* 71-84.

NOTES

[1] Evidence reveals that consciousness of some kind persists with the nonparticipation (e.g., because of lesions) of several brain regions, including the cerebellum, amygdala, basal ganglia, mammillary bodies, insula, and hippocampus (see review of literature in Morsella et al., 2016).

[2] See review with full citations in Morsella and Bargh (2011) and in Morsella et al. (2016). In brief, actions that are unconsciously-mediated can be observed during various kinds of unconscious states. For example, such actions are observed in some forms of coma/persistent vegetative states and epileptic seizures, in which *automatisms* arise while the patient appears to be unconscious. These unconscious automatisms include written and spoken (nonsense) utterances, complex motor acts, and other complicated actions. In neurological conditions in which a general consciousness is spared but actions are decoupled from consciousness, hands and arms carry out complex actions autonomously as in *alien hand syndrome*, and *utilization behavior syndrome*.

3 RIT effects arise in versions of the task that lack any kind of negative instruction (e.g., see the Baseline Condition in Allen et al., 2013). For example, Allen et al. (2016) instructed participants to hold in mind, as long as possible, one way of perceiving an ambiguous object (e.g., Necker cube). Involuntary "perceptual reversals," which involved involuntary entry into consciousness, occurred on around 80% of the trials.

www.ingramcontent.com/pod-product-compliance
Lightning Source LLC
Chambersburg PA
CBHW080518220526
45465CB00006B/2521